Ephraim Mosely

Teeth - Their Natural History

with the Physiology of the Human Mouth in Regard to Artificial Teeth

Ephraim Mosely

Teeth - Their Natural History
with the Physiology of the Human Mouth in Regard to Artificial Teeth

ISBN/EAN: 9783337130596

Printed in Europe, USA, Canada, Australia, Japan

Cover: Foto ©berggeist007 / pixelio.de

More available books at **www.hansebooks.com**

TEETH,

THEIR NATURAL HISTORY:

WITH THE

PHYSIOLOGY OF THE HUMAN MOUTH,

IN REGARD TO

ARTIFICIAL TEETH,

BY

EPHRAIM MOSELY,

SURGEON DENTIST,

9, GROSVENOR STREET, GROSVENOR SQUARE

LONDON:

ROBERT HARDWICKE, 192, PICCADILLY.

1862.

PREFACE.

No part of the human machine, comes so home to the regard and attention of all, as does the condition of the teeth. Our comfort, our health depend mainly, upon their correct and sound state. The loss or decay of teeth implies loss of health, consequent loss of appetite and good digestion. Indigestion must accompany any deterioration of their function.

The performance of this function of digestion properly can only be effected, and is a secondary process to a sufficient division of the particles of food, by effective trituration, and mastication in the mouth. Thus the food having already become reduced to a pulpy condition by admixture with the saliva during the process of mastication, is received into the stomach fit for digestion.

The food in this state of division, is in a condition to be acted upon by the juices secreted in the stomach, becoming forthwith converted into what is called chyme, the first state arrived at in the process of digestion. This process commences the series of functions of assimilation, and nutrition; without a correct performance of which, severe and painful ailments supervene.

The authority which long experience, and

great success in the improvement of the
mechanical department of the dental art, have
afforded us, will be our apology for offering on
this occasion, a few practical observations.
We shall aim in simple language to communi-
cate familiar and useful information to our
readers, divested of technical, or professional
phrases.

This we hope will prove acceptable, since
the subject of the defects of teeth, is of that
constant occurrence, and common place character, that all sympathize in it, and few will
refuse the advice we willingly afford.

The dislike which every one entertains,
to expose dental deficiencies is so great, that
we feel confident, our humble endeavour to
advance a few simple maxims, and to give
advice and confidence will be acceptable.

The good faith with which we offer this
advice, we have no doubt, will gain that confidence, and, with it the suffrages of our readers.
Inducing them to regard rules, which we shall
freely explain, thus securing with timely and
proper assistance, the enjoyment of a good set
of teeth.

On submitting the following practical
observations, with these objects and motives,
we solicit the candid perusal of our indulgent
friends the public.

9, *Grosvenor Street,*
      Grosvenor Square, W.

CONTENTS.

CHAPTER I.

THE VIRTUES, AND VICES OF TEETH.

ALL know more or less what tooth-ache is, we will therefore say but little upon this unpleasant subject. In this condition of suffering from tooth-ache, the unfortunate victim seeks relief and help, frequently in a reckless, and off hand manner. In haste to get rid of his urgent grievance, to the first druggist's shop he wends his way, having seen written up in his window, "Teeth Extracted," forthwith to part with a valuable member. This mutilation perhaps, being attended with other bad consequences. To the facts in the following anecdote we can bear personal testimony.

A baroness in her own right, being at her residence in a remote part of the country at a time when dentists were scarce, and medical men often disliked operating upon the teeth, was suddenly troubled with punishment from a molar in the upper jaw. She being a woman of strong character and determination, having been told that her blacksmith, residing in an adjoining village, was celebrated for fortune-telling, and tooth drawing, without hesitation sent for this country artist in horse shoes, and unravelling the mysteries of fate, to extract her tooth. This unkempt Mephistophilus, being ushered into the baroness's presence, pulled out a dirty handkerchief, with which to lap round the key tooth instrument, of his own forging. To this proceeding her ladyship demurred, and as a condition of submission to her tormentor, offered her own cambric handkerchief, which, from its fragrance and whiteness, sadly discomfitted this farrier artist. The man, nevertheless, with some difficulty and force, managed to extract the vicious molar.

This unfortunately, was the commencement of mischief. Pain being only a temporary grievance, is easily and quickly removable; but fracture from force, leaves permanent, and often irremediable mischief. The blacksmith's 'stithy' could forge horse shoes, and garden hoes, but turned out clumsily made key instruments for teeth. Thus his key was formed better to remove a granite boulder than a human tooth. This instrument having a heavy, powerful *fulcrum*, and equally clumsy *lever* to correspond, he necessarily fractured the baroness's alveolary process, and a portion of the jaw bone.

Her ladyship for the rest of her days, concealed in her mouth, the reminiscence of a blacksmith's skill in tooth drawing.

Let us above all things warn our readers against rash tooth drawing. Pain of the teeth, or of a tooth, we freely confess is a warning always to be regarded with suspicion, if not with fear, and apprehension. If under such circumstances, immediate application be made to a skilful and respectable dentist, relief will be prompt, and the source of mischief is easily removable.

This confirms universal experience, that preservation is a triumph, whereas mutilation is more or less a disgrace to art, either proving deficiency of skill in the artist, or want of resources and feebleness of art itself. Of the latter, we are always inclined to be sceptical.

In most instances the timely and judicious stopping of a tooth, which may be effected with gold, india rubber, as well as other materials, does all that is required, and preserves to the patient a valuable member. Stopping is a simple process, when practised by those who well understand its value, and çan accomplish cleverly all its intentions.

Let it be remembered how essential is a full mouth of teeth to the beauty of the human face. A white even set of teeth enhances the attraction, and completes the pleasing expression of the countenance.

With these advantages, the human face possesses a charm we cannot resist, and is a passport to our confidence, that we all acknowledge. The mouth opened, we are delighted with the laughing expression surrounding the lips. But our fascination becomes complete, when a perfect mouth gives utterance to speech.

Without teeth Demosthenes would not have been an orator, nor would his substituting sea pebbles have helped him, since without teeth they would have dropped from his mouth.

Thus endowed with perfect teeth, the mouth is a source of innumerable pleasures, and enjoyments;—seat of smiling beauty, fountain of persuasive eloquence. The converse of this picture is lamentable; by loss of teeth the gums recede, the face becomes shortened, and disfigured with premature wrinkles; thus, old age is caricatured by chin and nose approaching each other. Pronunciation becomes laboured, vocal harmony ceases, the voice also becomes inarticulate, lisping and broken.

The beautiful metaphorical language in which the teeth are apostrophized in holy writ, ought to plead on their behalf. In the "Canticles," the inspired writer, speaking of his love, says, "Thou art all fair, my love, there is no spot in thee. Thy teeth are like a flock of sheep that are even shorn, which came up from the washing; whereof, every one bears twins, and none is barren among them."

This eloquent description of teeth shows in what estimation they were held by the ancient, and sacred writers.

The science of physiognomy prefers some claim to speak on this occasion. To read the character, disposition, and temper in the features of the face, Lucian has said, is to carry the key, that unlocks Momus's little door in the breast.

To show up the man as a walking advertisement of himself, are pretensions not easily confuted. They are, moreover, accepted by the moralist, artist, and dramatist. Indeed creations established upon this basis, in imitation of nature, are exemplified in high art, being adopted by painters, and sculptors, and caricatured on the stage. Hogarth displayed a wonderful genius in this department of art, in depicting the familiar traits of human nature. Shakespeare, than whom no higher authority can be found, in his readings of the human character, says,

"There's no art to find the mind's construction in the face."—*Macbeth.*

But again, as though conscious of error, he qualifies this, in the following line,

"Your face, my Thane, is as a book where men may see strange matters."—*Ibid.*

Again—

"For, by his face, straight shall you know his heart."—
Richard III.

The influence of the human face is beautifully expressed in the following words—

"Vouchsafe to shew the sunshine of your face, that we, like savages, may worship it."—*Love's Labour Lost.*

But a word is to be said on the other side. Coleridge relates a ludicrous *denouement* of the fallacy of face. At table one day, watching a silent thoughtful looking intelligent face, from which he promised himself a treat, Coleridge was somewhat surprised to hear, when a dish of dumplings was served up, the exclamation, "Them's, the jockies for me!"

The face of the man who had made a vow, that at the child's play of "cup and ball," he would catch the ball on the small end, 100,000 times, had acquired the severity, and fixedness of Newton's.

Teeth! without which "the human face divine" is a moping, melancholy mask—the speech, a mumbling mockery—the living manly form—a monumental mummy, meagre—mortified—moribund.—

"So coldly sweet, so deadly fair,
We start! for soul is wanting there."—*Byron.*

"Our virtues would be proud, if our faults whipp'd them not."—*All's well, that ends well.*

The correlation between virtue, and vice, we view in the abstract as an unfortunate, although inevitable condition of human nature. Where this kindred begins, or where ends, it is not our task to say, or to give their traditional genealogy, which is as ancient as man. The ear

reluctantly acknowledges this discordant anti-
thesis, as reason repels the moral paradox: but,

"The Gods of our pleasant vices, make instruments to
scourge us."—*Cymbeline.*

Teeth, the armed ægis of life! Like the
shield given to Pallas, which, when Medusa's
head was placed on it, turned all who viewed
it, into stones: so teeth, before the breath of
life has passed over the soft velvet birth-place,
where they lie nascent in serried ranks, spring
from their embryonic beds, to be converted into
crystal stones.

Of what are teeth formed—of such mar-
vellous structure—of such uniform symmetry—
shape, and design? This marshalled armoury
consists of three materials, technically called
dentine, cementum, and *enamel;* added to which
is the pulp, or nerve bulb. The first of these
materials, *dentine,* constitutes the substance of
the tooth; the second, *cementum,* the substance
of the bone; and the third *enamel,* a pris-
matic deposit, imitated only if successful by
that glaze, which at Sèvres is given to the
primest *bisquet* porcelain.

The numerous inconveniences sustained by
victims of the decay, and vices of teeth,
would be an ungracious catalogue to relate,
or reveal. The sufferer would rather avoid a
repetition of his griefs. The possessor of
virtuous teeth, would turn away with aversion
and disgust from the miserable history.

Teeth! the simplest organized structure in
the whole animal machine—the lowest form of
growth and development. Nevertheless have not
the philosophic discoveries of Cuvier, Blumenbach,
and Owen, found in the simple tooth, a guide
post in natural history, and a key-note in

science? Have they not discovered in it, the
type of species? In this simple osseous
structure, these eminent naturalists have laid
the foundation of a new science much regarded
at the present time, namely the science of
Palœontology.

Mouth! domicile of teeth: too apt to be
associated with gross sensual ideas. Mouth!
the source of innumerable enjoyments and
delights to man—seat of smiling beauty—of
intellect—of comeliness—fountain of God-like
eloquence—from whence in curling lines flow
expression of sentiments of love, of delight
of maternity, of infantine ecstacy, and of
affection!

Baron Von Humboldt says, "Man is man
only by means of speech, but in order to
invent speech, he must be already man."
"Language also, like amber, embalms and
preserves the relics of ancient wisdom."

Homer, through his hero warrior, replied,
two thousand years ago to Talleyrand's
Machiavellian proverb, which says. that
"language was given to man, to conceal his
thoughts."

Pope has rendered the justly earned cas-
tigation thus—

"Who dares think one thing, and another tell,
My soul detests him as the gates of hell."

Words rule the world. Without their teeth
what would D'Israeli, or Gladstone be in the
House of Commons; Lyndhurst, and Derby in
the Peers. Shakespeare makes Henry VIth
exclaim—

"My mouth, shall be the Parliament of England."

Words and language, are the public orators

of thought. Thoughts lie nascent in the brain to be awoke by words, revealing the precepts of reason, religion, and morality.

A blind man can tell a cultivated intellectual person by the flexibility, cadence, and intonation of his voice. The blind man's attention becomes fixed—overwhelmed with emotions of pleasure, he listens to the harmonious ringing vocal music, wrapped in admiration and delight. This vocal music is produced by teeth.

Language, is a miracle, but utterance of words is a still greater one. It is the prerogative of man only, to inherit the gifts of language and speech, demonstrating this miracle.

Mouth! through which is sustained life. Without taste the provocative of appetite, the high arched palate would become a deserted highway; if teeth did not cater to taste, whose function is thus seated in the mouth—life would cease. What is life? question ever asked—never answered—mystery from all time, as far, as ever from solution. Yet without mouth and teeth, this problem would arrive at a negative solution, since life would become extinct, and in this way the mystery becomes negatively solved.

Art, has succeeded in imitating by artificial teeth the mechanism necessary to mastication, so as to fulfil the function of deglutition—thus, preparing food for the nourishment of man, to continue and prepetuate this mystery of life.

"Virtute vicit incommodum naturæ."
Cicero de Fin.

Nevertheless art cannot imitate the mechanism of voice, or accomplish the function of speech.

Sound—governed by vibrations, reverberations, and modulations, has a mechanism of infinite beauty, and perfection in the human instrument. This mechanism consists of the functions of the pharynx, larynx, trachea, with their appendages, namely, the epiglottis, and vocal chords; contributing together a most elaborate musical instrument — an instrument soon put out of tune; so delicately constructed, as to be easily spoiled. Let the tongue, the teeth, palate, or nose become altered, or diseased, the uvula elongated, or tonsils enlarged, and hypertrophied, the harmonious music of voice thenceforth becomes destroyed.

Art, steps in to effectually alleviate the mischief — *staphyloraphy*, or *tonsillotomy*, and sometimes *tracheotomy*, save voice, and life, and give to the artist, the ardent, well earned gratitude of his patient.

Freedom of judgment, the load star of human intellect—the high prerogative of reason, proclaims the triumphs of art, on all occasions, to be most excellent, to subdue and convert nature.

" Nihil est arte, quod non prius in naturâ."

Cultivated art, approaches nearest to nature in its endeavours to satisfy human wants. Art is the reflection of nature in exactness, and precision.

" Ars dux certior quam naturâ."
Ac. de Fin.

Objects in nature and art, which observe periodicity, and order, forcibly appeal to the imagination. Thus the sun, the moon, the seasons, day and night, equally with a clock, a watch, or a steam engine, inspire the

B

10

wonder, and almost adoration of the savage. Herein is established the ascendancy of nature, and art. Let us hold faith in nature, science, and art, to sustain, and foster the undying hope, which throws its forward shadow into eternity.

We may, in the words of Quintilian, say—

"Ars est, quœ disciplinâ, percepi potest."

All the senses centre in the mouth. The mouth is a well defended garrison; the forces of speech, of taste, of hearing, and of smell are contained in its store rooms, and magazines. Sight, holding watch, and ward over its sacred stores, is seated above, with eaglet gaze from its battlements.

It is in search of the useful, in the struggles for material supremacy, we wear out our energies in the present day. We have no time allowed to indulge, or recreate in fanciful speculation; stern realities possess our thoughts, and inspire resolution to accomplish our endeavours. Sincere in our search of truth, we acknowledge her severe but benign influence, and worship her with implicit veneration. Thus, in our speculations and schemes, study, thought, and reflection, flow in a healthful stream. Error, when its existence is practically demonstrated, and publicly announced, although tardily, and sometimes reluctantly confessed, is manfully atoned for.

This reality governing our actions has taught us to conquer inert, dull, opposing matter, to annihilate space, and time; has enabled us to say, "the lightning bears my messages, and the air is my courier." Thus, amid such influences the progeny of our intellect is vigorous, hale, and robust, we have no ricketty, weedy, sentimental creations appealing for sympathy. Our subjugation of material matter, of the external world, has substantially realized the Golden age, and by humanizing tendencies, given a poetry, to our practice.

We are essentially a practical age, even our poetry and fiction deal in familiar realities, for "truth is greater than fiction." Laborious triflers cannot be heard. He who has a word to deliver, must speak out, or give way. The genius of the age, like Minerva, springs out armed, for action. Nevertheless it is a just age. We have no "Martyrs of Science." Persecution of opinion, has passed away. The *summum bonum*, is the "greatest happiness, of the greatest number."

The present age aims at every thing which human invention can accomplish. While creating mighty inventions to influence the destinies of nations, its genius stoops gently to gather up a nail, or a feather. Like one of its mighty engines, it will crush a rock, as also it will head a pin. Thus, the age in which we live, is great, because it is just, and considers nothing mean that is useful.

Invention is one of the great marks—is the symbol of genius, or may we not say that genius is invention expressed by "a sharpening of the spiritual sight to discern hidden aptitudes." He who has the greatest store of intellectual material, has the greatest means of invention. "Divine honours," remarks Bacon, "were rightly by antiquity, attributed to inventors, but upon such as had deserved well of their country, only heroical honours were conferred. Without any blind or servile reverence of antiquity, yet while vaunting the triumphs of modern science, and art, we must remember the difference between invention, and improvement; the cultivation of acquired knowledge, and the labour and difficulty of emerging from ignorance, and inexperience.

In our day, not to travel back a century earlier, to the discoveries of Newton, and Harvey, and the invention of the Stocking Frame, we claim our men of science in Hunter, Watt, Davy, Dalton, Faraday, and Owen. Our men of machinery, and successful enterprise resulting from machinery, record amongst their ranks, the names of Bridgewater, inventor of Canals; Watt, of the steam engine; Arkwright, of spinning Jennies; Heathcote of lace machines, and Stevenson of railroads.

On naming these, we must not pass by the first Sir Robert Peel, distinguished, not only in successful enterprise, but by founding a noble house, and an original stock of statesmen. The first Sir Robert was himself also a legislator of no mean capacity. It has been the lot of few, from being the sire of so great a son, to achieve a fame more lasting, or of a nature more genial.

If, for a moment we pause to consider the physiological incidents depending upon teeth, we shall be convinced how nearly allied is our personal comfort, health, and enjoyment to their healthful condition. The orator would become worse than dumb, by the loss of teeth, since if he used his mouth to give utterance even to beautiful ideas, clothed in vigorous, and glowing language he would assuredly lose his fame.

CHAPTER II.

MECHANISM OF THE MOUTH.

We have already scanned the "virtues and vices" of teeth. Let us therefore take a peep into the human mouth, curiously review its mechanism, much as we would the machinery of *a watermill*, or *windmill*, or any other useful mechanical invention.

It is on such occasions, and by such comparisons, we have to acknowledge the superiority of nature, over art. We find in all mechanical contrivances, whether natural, or artificial, harmony in the economy, with fitness, and propriety, in adaptation of means, to an end; especially observed in imitations of nature, by achievements in art. This is discovered in the human mouth. The same rule, applies to all inventions, however consummate, may be their manifestation of skill, and ingenuity, so that we can trace the original type, to some mechanical development in nature.

Thus we find the mouth, supplied with a singularly simple, yet beautiful apparatus, for the performance of its functions, of preparing food, for the human stomach, for preservation of human life. The teeth are divided to perform the office of mastication into two sets. The molars, or as they are significantly called, the *grinders*, (for by that name they sufficiently bespeak their function,) and insisors, which name will also explain the office they perform in the mouth. The incisors occupy the front of the mouth; these constituting the two sets of teeth we have named. The molars, and bicuspids, fill up the lateral and posterior parts of both sides of the mouth. Thus, we find the human mouth represents the type of a watermill in miniature.

The cornmill simply grinds between the faces of two millstones, prepared for the purpose, the grains of corn, into flour. The mouth through the agency of the teeth, operates upon materials of various degrees of hardness compactness, and density. These materials are converted into a pulp, by the moisture of the salivary secretion, during the trituration of the teeth.

Thus water propels the wheels, by which the stones grind grains of corn; and moisture in the salival liquid, is necessary to the proper division, of particles of food, in this natural watermill, the mouth.

But the mouth is a more complicated machine, than a watermill. Corn, is submitted in its natural whole state, of distinct grains, to be triturated, until ground into flour. The food, before being admitted into the portico of the mouth, undergoes scrutiny, being presented to it in various forms, sizes, and conditions. If it be too large, the front incisors, divide it into a suitable size, for its easy trituration, or mastication, by the proper millstones of the mouth, namely. the molar teeth. In biting, or separating a large morsel of food, for its easy admission into the mouth, the bicuspids are found extremely serviceable. They with the incisors, make a clean division of the morsel to be received, for the grinders to manducate.

It is a mistaken notion entertained, that teeth are only serviceable, for the division and mastication of food, for the stomach. In the human mouth, teeth subserve also, to the important function of reverberators, or vocal keys, for conveying, and modulating sound.

The human teeth are, moreover great ornaments to the human face. We are sorry to say, that they are far too often appreciated, only as ornaments, a mistake nevertheless prolific of profit to the dentist. We can safely affirm, that where a patient consults his dentist, upon the economy of his mouth, with regard to the serviceableness, and utility of his teeth, he is much more often prompted to do so, by influence of vanity, and regard for appearance. It is a great mistake, to consider artificial teeth, as merely ornamental, which many of the best clients of the dentist, certainly do. On the contrary, when artificial teeth, are skilfully and properly made, and fitted, they are invaluable to the wearer, for comfort, and utility, as well as ornament.

Mastication can be as well completed, by artificial, as by natural teeth; speech, moreover, depending upon their valuable assistance. With this erroneous impression. that the principal advantage of artificial teeth, is ornament, many persons, so long as they retain their front teeth, unfortunately disregard the loss, of their more useful molars, or back teeth.

The maxillary mechanism, is constituted of many muscles, and large bones. The muscles, according to their economy, as in all other instances, act by antagonism, being consequently, often much at rest.

The two large maxillary bones, like the famed, and classical "pillars of Hercules," guard the *faucial isthmus*. When the muscles are passive, at rest, the teeth support the immense superior maxillary wall, thus preserving the contour, and symmetry of the face. This can only be properly effected, by the integrity of the molar teeth, situated behind the incisors. The incisors, in front of the mouth, being thus preserved from injurious pressure, and contact.

The molars, at the back part of the mouth, during the process of manducation, prevent the front teeth, from coming in contact, and by so doing, prevent also the front incisors, from being damaged, by attrition, or friction, which would have a tendency to loosen them. It should be always remembered, that the incisors, are only subservient to cutting food in parts, having nothing whatever to do, with the function of mastication; this function of mastication, being performed exclusively by the molar teeth, or grinders.

Hence, is found the imperative necessity, of supplying lost molars, directly the dismemberment occurs, to save the front teeth, from the inevitable collision, with each other, to which we have referred, such collision having the effect, of wearing them away, or loosening them.

Moreover we must consider the incisors, as wedges; they are wedge-shaped, the top incisors also overlapping, the bottom front teeth. By the loss of a molar, or molars, these wedges, assume a different line of direction, from the natural perpendicular, thus, having a tendency, to force each other from their sockets.

We must again observe, since it cannot
be too much regarded, that the molars, only
are essential to mastication. The machinery of
mastication, in the human mouth, is exclusively
performed by the molar teeth; the incisors, or
front teeth, being useless, and improper for that
purpose. The incisors, are the *cutting*, as the
molars,.are the *grinding*, parts of the mill machine,
of the human mouth.

Thus, the maxillary wall of the mouth, is
weakened by loss of even a single molar tooth,
the security of the remainder being thereby
disturbed, and endangered. As with a file of
soldiers, each supports his next comrade, so with
the file of molar teeth, one removed, leaves a
breach, which weakens the whole rank.

Again, the teeth in the opposing jaw, being
deprived of the healthful support, during mastication, to which they have been accustomed,
and which nature requires, in the absence of
this support, become loose in their sockets,
speedily falling away. Moreover, loss of the
molars in one jaw, deprives the function of the
opposite molars in the other jaw; consequently
to avoid further mischief, they should be imme-
diately supplied, by artificial teeth.

The mischief results in this way, loss of
the antagonistic support, of the opposite teeth,
to which we have just referred, gives a relax-
ing tendency to their opponents, which consequently induces disease in the floor of the
socket, which is termed *exostosis*, or a deposi-
tion of bony matter, forcing the teeth, gradually
to emerge from the sockets. Thus the bony
portion of the teeth, not enveloped with enamel,
becomes exposed to external influences, for which
nature did not intend it, namely, the influence

of atmospherical changes, the action of condiments, and other stimulants or excitants, admitted into the mouth ; and ultimately the teeth drop away from the sockets without any sign of decay.

By regard to the rules, and practical cautions, we have offered, mastication is perfectly carried on. The front teeth are also preserved, conferring, the enjoyment of health, and frequently the extension of life, to individuals.

CHAPTER III.

The teeth are the simplest organized structure, the lowest form of growth, and development, in the whole animal machine. They have been typified by Dr. PALEY, as a manifestation of design, showing provision for future wants. The incomplete development of the teeth at birth, shows a provident contrivance for a future state of development in the animal economy, and of design in creation. Paley says — " That intelligence, which was employed in creation, looked beyond the first year of the infant life, yet while she was providing for functions which were after that time to become necessary, she was careful not to incommode those which preceded them."......... What renders it more probable that this is the effect of *design*, is that the teeth are imperfect whilst all the other parts of the mouth are perfect. The lips are perfect, the tongue is perfect; the cheeks, the jaws, the palate, the pharynx, the larynx are all perfect: the teeth alone are not so. This is the fact with respect to the human mouth. The fact is also, that the parts above enumerated, are called into use from the beginning of existence, whereas the teeth would be only so many obstacles and annoyances, if they were there.

Again, speaking of the creation, " When a contrary order is necessary, a contrary order prevails; thus, in the worm, the teeth are the first things which arrive at perfection. As the worm is hatched from the egg, the insect begins to gnaw as soon as it escapes from the shell, though all other parts be only gradually advancing to maturity." But, "............ teeth are formed in the gums, *there they stop;* the fact being that this further advance to maturity would not only be useless to the new born infant, but extremely in its way; for the act of sucking by which it is for some time to be nourished, will be performed with more ease, both to the nurse and the infant whilst the inside of the mouth, and edges of the gums are smooth and soft, than if set with hard and pointed bones. By the time they are wanted they are ready."

Again, ... " In respect of the human teeth, prospective contrivance looks still further. A succession of crops are provided, and provided from the beginning. A second tier of teeth being originally formed beneath the first, which do not come into use for several years afterwards. This double or supplementary provision, meets with a difficulty in the mechanism of the mouth, which would appear almost insurmountable. The expansion of the jaw, in consequence of the growth of the animal, and of its skull, necessarily separates the first set of teeth from one another, however compactly composed they may be."

This would be very inconvenient. In due time therefore, when the jaw has attained a great part of its dimensions, a new set of teeth, loosening and pushing aside the old ones, springs up. These are more exactly adapted to the space which they are to occupy. Rising also in such close ranks, they allow for any extension of line, which the subsequent enlargement of the head, may occasion." (*)

A tooth is anatomically divided into four parts, namely:—the crown, which is seen above the gum, invested with enamel; the neck, is that part surrounded by the gum, between the crown, and the alveolus, it is next the root, or fang of the tooth, hidden from view, being seated in the alveolus: lastly, a cavity containing the pulp, which is a soft, vascular, very sensitive substance.

The teeth are fixed in their sockets in the alveolar ridge, by an articulation peculiar to the alveolus, no other such articulation existing in the human body. It resembles the fixture of a nail driven into a board; the teeth being kept firm, equally by the closeness of their fit, as by the adhesion of the *periosteum* lining the sockets of the teeth, and investing the fangs, as also by the gums. This singular articulation of the teeth in the alveolus, is called *gomphosis*.

The adult teeth are 32 in number; viz., *incisors, canines, bicuspides*; (these are peculiar to second dentition, or permanent teeth) and *molars*; the first and second of which are the largest of all the teeth. The *dens sapientia*, or last molar, is the last tooth in the mouth. This tooth is not so large as the others, the roots being united into one. The

(*) Natural Theol. chap. xiv.

dental arch thus formed by the line of teeth
is a *parabola*. The crowns of the canine,
slightly deviate from the parabolic line. When
the mouth is shut, it is worthy of remark,
that the teeth do not close on each other in
exact junction, tooth for tooth, since the upper
slightly overlap the under.

The *structures* of a tooth are four in
number. These may be seen with the assistance of the microscope. They consist of, 1st,
The *dentine;* this forms the great mass or
mould of the tooth. 2nd. The *enamel;* constituting the exterior coat of the crown. 3rd.
The *cementum*, or *crusta petrosa;* this forms
a thin layer around the root, not including
the foramen, where vessels and nerves enter,
and is traced over the enamel. 4th. The
pulp, which is found in the central cavity of
the dentine. These tissues have each distinct,
and peculiar characters. *Dentine*, composing the
great bulk of the tooth, assimilates in its
character to the elephant tusk, or ivory of
commerce. *Cementum*, corresponds with the
osseous tissue of the body generally, is
the most highly organized, and vascular of
the dental tissues, and is the last formed.
Enamel, is a semi-vitreous looking substance,
of the colour of pearl, with glossy smooth
surface, being also composed of solid prisms,
set side by side, on the dentine of the crown,
in a direct perpendicular line. The prisms
have a wavy course, their curves being parallel.
The enamel becomes thinner at the neck of
the tooth, around which it terminates.

According to Liebig, enamel contains no
animal matter. Berzelius gave only 1 per
cent, it is the hardest of all animal substances.

Mr. Goodsir's researches, establish that the development of the tooth, its *nisus-formœtivus*, is referrible to the same kind of germ, as is that of a hair, or feather, namely, a *bulb*. The first appearance of the pulp of a tooth, occurs about the seventh week of fœtal life; it is the germ of the upper anterior molar, the first tooth that appears. Thus in its progress dentition commencing in the upper jaw, fine tubes radiate horizontally from the pulp cavity, into which they open, to the surface of the tooth, in an undulating course.

The first appearance of the pulp of the tooth, is in the form of a minute egg-shaped *papilla*, arising from the bottom of a groove behind the edge of the jaw, in the mucous membrane of the mouth. The substance of the tooth when it commences formation, is produced in the following order:—1. *Dentine*, is developed from the pulp. 2. *Enamel*, from a granular substance deposited between the capsule and the pulp. 3. With respect to the *cementum*, nothing positive is known, but it is most probably produced from the periosteum.

Nasmyth has shown that the papilla, or pulp, is cellular. By its cells the dentine tubules are produced; the pulp becoming hardened by earthy matter, gradually deposited between these cells, layer after layer, from without inwards; thus, all becomes solid, except the foramen of the pulp.

The formation of the *permanent* teeth, originates from the mucous membrane, in the same manner as the milk, or deciduous teeth originate, but they are independent, and have no connection with them.

The periods of the eruption, or development of the 20 deciduous, or milk teeth, are variable. The central incisors appear from 5, to 8 months, after birth; the lateral incisor from 7, to 10; the anterior molars, from 12, to 16 months; the canines, from 14, to 20; and the second molars, from 20, to 36 months. The teeth of the upper, and lower jaws appearing nearly contemporaneously. The advent of second dentition, or of the permanent teeth, begins usually about 6, or 7 years after birth, but is variable as to precise time; the previous crop of deciduous or temporary teeth, being shed in consequence of absorption of their roots, in the same order in which they first appeared.

These deciduous milk teeth, are smaller than the permanent, and not of such firm texture, although the milk *molars*, are larger than their successors, the permanent, bicuspides.

The Periosteum, which teeth possess in common with the bones of the body, serves to display, situated around the teeth, an instance of what Paley has spoken of as "interrupted analogies," frequently manifested in the mechanism of the anatomy of animals. An instance of deviation from the usual course of organization, is manifested for obvious uses in the epidermis at the finger ends, giving way to nails, to assist with the faculty of tentation, the important tactile function. Had so exquisitely sensitive a membrane as the periosteum invested the teeth, as it does every bone of the body, their necessary action, exposure and irritation, would in consequence have subjected the animal to continual pain. General as is the periosteum over all the bones of the body, it is not the sort of covering, or integument which suited the teeth; what the teeth stood in need of, was a strong, hard, *insensible* defensive coat : exactly such a covering is given to them, in the ivory enamel, which adheres to their surface.

Differences in the formation, and character of teeth exist in man, from those found amongst other animals.

The human teeth present in their perfect, and adult growth a close, compact, uninterrupted series. Thus they differ in man from all other animals, of which we will give a few examples. The *Chimpanzè*, or *Orang*, which approaches nearest to man, has pairs of crescentic alveolar ridges the convexity of which is *outwards* in the lower, and inwards in the upper jaw. The molars of the Indian and African elephants, differ also from the molars of man. Those of the elephant are complex; thus each tooth of the elephant is as if made up of a certain number of distinct teeth, bound up into one mass. Man is omnivorous; and right, or wrong, will eat what he likes to the end of time, " so convenient is it," says Franklin, " to be a reasonable creature."

Professor Owen says, " Though the human cranium, varies in its relative proportions to the face in different tribes, according to the degrees of civilization, and cerebral development which they attain, and though in the more debased Ethiopian varieties, and Papuans, the skull makes some approximation to the *quadrumanus* proportions, still in these cases, as well as in artificial, or congenital distortions of the skull, it is always accompanied by a form of the jaws, and by the disposition and proportion of the teeth, which afford unfailing, and impassible generic distinctions between man and the ape."

But man manifests great odontolgic oddities,

C

which has given rise to the aphorism, "Man, differs more from man, than man, from beast." "*Simia quam similis, turpissima bestia nobis.*" Shakespeare rebukes the hunchback, king Richard the third, who has had a fair share of contumely clapped upon his hump.

> "Teeth hadst thou in thy head when thou
>   was't born,
> To signify thou cam'st to bite the world."
>   *Richard III.* .

Children have been born with a complete set of teeth in their mouths. Pliny mentions the singular conformation of the teeth in Pyrrhus, the Epirote King, in whose jaws the teeth were all united by the crowns. "The younger Pliny states, that the renowned Marcus Curius, Consul of the Roman republic, 270 years before the Christian Era, had a full set of teeth at birth. Thus was he named *Dentatus.* The same author mentions the case of Papyrius, and of a lady named Valeria, who had all her teeth at birth.

Zoroaster, the Persian legislator, is also reported to have had all his teeth at birth. The old historians, from whom Weinrich took the facts, probably thought Zoroaster was destined to become the exponent of wisdom and morals, from being so extraordinarily and precociously gifted.

Louis XIV. of France, whom some writers call great, because he lived contemporaneously with some great men of his country, was born with two teeth; as was also his Secretary of State, Cardinal Mazarin. The celebrated Grotius, who then lived in France, prophesied that the royal baby would prove a dangerous character; and that like the nipples of his wet nurse, bleeding, and torn by the voracious infant, neighbouring sovereigns, would be subjected to the depredations, and robberies of the future king. Scottus, in his physiological curiosities, relates from a report of Nurenberg's, the case of a Spanish dwarf who had all his teeth when born, and never lost one of them—got a beard in his seventh year—and had a son in his tenth. (*)

Cutting of teeth in old age is a curious, though not uncommon phenomenon. Thus it would seem, that occasionally nature in her prodigality, makes an effort to reverse her laws, by attempting to repair the defects of age. The *grand climacteric*, so called, having passed, " sans teeth, sans eyes, sans taste, sans everything," to hail the advent of second childhood a third dentition has occurred. The process is accompanied with much pain from the naturally callous, and hardened state of the gums, at this period of life. John Hunter states, that he witnessed the reproduction of a complete set of teeth in both jaws. These teeth must have been cruelly in the way—certainly more injurious than serviceable. Without a renewal of the sockets, which under such circumstances, have been long absorbed, these teeth, would necessarily be irregularly developed, not having proper fangs. Thus they would interfere with the vicarious office of teeth, which the gums have generally, at this time of life acquired.

John Hunter frequently transposed teeth, in conjunction with a celebrated dentist of

(*) Lecture by Dr Jacobi, published in the " American Medical Times," which gives many other physiological curiosities of a similar kind, upon the subject of teeth.

that day, who lived in Soho Square, of
the name of Spence, a clever and intelligent
man; but unpleasant consequences resulting, the
practice was relinquished, not without a little
scandal attending the procedure. He also transplanted a human tooth into the cockscomb, with
success, as the preparation to be seen in the
Museum of the College of Surgeons testifies.
The tooth thus transplanted, formed adhesion
by vascular union with that ornament of the
bird's head.

CHAPTER IV.

Let us pause for a moment to take a retrospect of the views of the ancients, upon the subject of teeth.

Art in any but its infant state," says Mill, "pre-supposes scientific knowledge: several sciences, are often necessary to form the groundwork of a single art.

" Pater ipse colendi
Haud facilem esse viam voluit ; primus que per artem
Movit agros ; curis acuens mortalis corda."—*Virgil.*

The dental art is no doubt as ancient as Adam; a Hindoo, or a Chinaman, would doubtless carry its antiquity to a still greater remoteness. The ancients bestowed some regard to this art, although it was of a confused, mechanical, unskilled kind. In the present day we enjoy a peculiar privilege, namely, the contemplation of two types of civilization, the *most ancient*, and the *most modern.* But modern civilization can trace little, or no relation with the most ancient. The colonies founded by the Egyptians, namely, Phœnicia, Tyre, and Sidon; and Carthage, by the Tyrians, on the northern and southern coasts of the Mediterranean, first awakened the early Greeks from barbarism to civilization. The Persians, Egyptians, and western Asiatics formed the link which connected the fabric of modern civilization amongst (if we may so speak) the *modern ancients,* of Greece and Italy.

We are becoming daily more able to compare ancient, with modern civilization. Civilization, first planted its footmarks in the far east, slowly to wend its devious course westward. The extreme limit of its eastern boundary, has until within the last century been so little known as to be almost considered a fabulous world. The shores of the Indus, the Ganges, and the Jumna; the Chinese, Mogul, and Tartar Empires, have only become recently invaded by modern civilization. Even the Romans and Greeks, notwithstanding their enterprise, scarcely heard of their more ancient eastern rivals in art, and science. The civilization of the Hindoos, Chinese, and Persians, arrived at a fixed point, beyond which it made no further progress, or withered under the searing influence of despotism.

On the fall of Constantinople, the Greeks, who took refuge in Italy in the 15th century, were principally monks. They brought with them their books containing the science and literature of antiquity, and instructed the Italians how to read them. If they had not Europe must have continued to this time in a state but little removed from the barbarism of the dark ages. As we look back with delight, to the revival of this epoch of letters and taste, admiring the researches, and elegant labours of Petrarch, and Boccaccio, we shall remember with gratitude what is owing to the influence and teachings of Theodore Gaza, George of Trebizonde, Chalcondyles, and other learned Greeks, who elevated Florence, to the glory and renown of a second Athens.

The dwarf on the giant's back, sees further than the giant, we should therefore regard with reverence and gentleness the labours of antiquity : of those

"Immortal spirits in whose deathless song,
Latium and Athens, yet their reign prolong,
And from their thrones of fame, and empire hurl'd,
Still sway the sceptre of the mental world."—*Macaulay*.

Lord Brougham beautifully says—

"The globe itself as well as the science of its inhabitants, has been explored according to the law which forbids a sudden and rapid leaping forward, and decrees that each successive step, prepared by the last shall facilitate the next. Even Columbus followed several successful discoverers on a smaller scale; and is by some believed to have had, unknown to him, a predecessor in the great exploit by which he pierced the night of ages, and unfolded a new world to the eyes of the old.

The arts afford no exception to the general law. Demosthenes had eminent forerunners, Pericles the last of them. Homer must have had predecessors of great merit, though doubtless so far surpassed by him, as Fra Bertolomeo, and Pietro Perugino, were by Michael Angelo and Raphael. Dante owed much to Virgil; he may be allowed to have owed through his Latin Mentor, not a little to the old Grecian; and Milton had both the Orators and the Poets of the ancient world, for his predecessors, and

his masters. The art of war itself is no
exception to the rule. The plan of bringing
an overpowering force to bear on a given
point, had been tried occasionally before
Frederic II. reduced it to a system, and the
Wellingtons, and Napoleons of our day, made
it the foundation of their strategy, as it had
also been previously the mainspring of our
naval tactics."*

To our knowledge of the art of printing,
of which the Greeks and Latins were ignorant,
we are indebted, for the high station we hold
at present ; this circumstance would of itself
cease to create wonder, at the rapid strides
and improvements continually made, in consequence of this discovery of the art of printing.

We regard with admiration the success,
and superiority of the ancients in the developement of the body, cultivation of personal
perfection, and charms. The exercises of
the Gymnasia, and the Pálæstra, brought the
corporeal powers to the utmost capacity for exer-
tion of muscle, display of manly vigour, and
graceful carriage. In addition the varied, and
perhaps almost too luxuriant use of the bath,
with its appliances ; the minute and scrupulous
attention to the nails, hair, teeth, and skin ;
their use of fragrant perfumes, unguents, oils,
essences, and flowers, proclaimed high civilization. These bespoke cultivated refinement, self-
esteem and respect, which no doubt influenced
the inward consciousness, enlarging the appreciation of the beautiful, and sublime in nature, and
the imaginative in art.

---

* "Tracts, Mathematical and Physical," pp. 282-3,
by Lord Brougham, L.L.D , F.R.S., &c., &c., &c

We are behind the *modern* ancients in this respect, as we have reason to believe we are also behind the earlier ancients, since in the far east, where we have recently penetrated, we have discovered that the latter, practise in the present day, all the appliances of exquisite luxury, cultivated refinement, and art.

In our day it would be unjust, and libellous to say with Pope

" Though Artemisia talks by fits,
Of councils, classics, fathers, wits,
Reads Malbranche, Boyle, and Locke ;
'Twere well if she would pare her nails,
And wear a cleaner smock."—*Pope.*

From what Herodotus says, speaking of the ancient Egyptians, the custom existed at that time which prevails in our day, of dividing the practice of the healing art, into *specialties.* *Professors of surgery* and medicine practised in several distinct departments, one for the eyes, another the head, and inward diseases, and a *Dentist for the Teeth.*

Aretœus the Cappadocian fairly confesses, that the cause of tooth ache, is known only to God.

Albucasis at the end of the thirteenth century, first taught the use of artificial teeth. In " Al Tarif," or the Art of Healing, he gives figures of several dental instruments then in use. Galen in the second century, described the milk teeth ; also the operation of taking out stumps with a forceps-like instrument.

Belzoni has established the fact that the ancients were acquainted with the art of making artificial teeth, having found artificial teeth in their catacombs, and tombs. Surgical instruments for operating on the teeth, have been also discovered amongst the ruins of Pompeii, and are now deposited in the British Museum. Lucian introduces us to the toilet of a Greek belle, surrounded with rouge, cosmetics, and *tooth powder*.

Athenæus describes the ladies of his day as playing off their charms much after the fashion of the modern fair ones of London. or Paris; in his " *Deipnosophista, &c.,* " Lib. xiii., he says, speaking of a handsome woman :—" She had beautiful white teeth, so she was laughing all the time, that the company might see her pretty mouth."

The pure and poetic Catullus, who flourished 40 years before the Christian Era, ridiculed the foppery of his detestable, and mercenary rival Egnatius. This wretch committed all sorts of crimes for money, yet, *cultivated his teeth.*

That extraordinary man Saunderson, the blind mathematician at Cambridge, in company. one evening with a distinguished blue-stocking, to whom he had been introduced, but who had just left the room, was expatiating on the brilliant white teeth she possessed. The company expressing their astonishment at the truth, and accuracy of his encomium, enquired " how could he know that ?" " I have reason to pronounce," replied he, " from her conversation, that she is a woman of superior mental culture, but I cannot account for her constant laughter, except that she has a fine set of teeth, and likes to show them."

The peculiar and distinctive conformation exhibited amongst the teeth of different species of the animal kingdom, have rendered them if not the sole, the safest guide to determine the several species, and varieties of extinct animals. Thus they have formed the guide, to determine the several types of the animal creation, revealed to us in the researches of geology, and in the specimens of fossil organic remains, which are constantly brought to light. The character of the bone, or bones of a limb, or a tooth, or even a portion of them, determine the whole formation of the animal. They also constitute an index of the habits, mode of existence, character of food, haunts, size, and every other distinction peculiar to those animals which lived upon the earth, and in the waters, before the creation of man.

Cuvier says, "I found myself placed in a charnel house. At the voice of comparative anatomy, every bone, and fragment of a bone, renewed its place! I cannot find words to express the pleasure I experienced, when I discovered one character, seeing as I did, how all the consequences I had predicted from it, were successively confirmed."

Thus Professor Owen, discovered the gigantic fossil *Megatherium*, of South America, from the evidence afforded him in examining the formation of the fragment of a tooth.

This difference in the structure of teeth, affords the naturalist, the means of discovering to which group or species of animals, any particular formation of tooth may belong. The formative arrangement being always invariable, and constant for each individual group. The portion of fossil tooth referred to, which guided Professor Owen in his scientific paleontological investigations, belonged to the family of the *Edentata*. Upon minute examination of the structure of this fragment, he found an exact analogy existing between it, and the sloth species, living at the present time. Some curious facts had to be elicited, to confirm the truth of Professor Owen's proposition. Thus he ascertained with precision, that it was a vegetable crushing tooth, not sufficiently dense or hard in structure to grind down roots, nevertheless, fitted to crush leaves, and young shoots.

Leaves, and young shrubs, constitute the food of the modern sloth. The system of teeth of the common sloth of our time is very simple. It has very small canine, and no incisor teeth. The mouth consists of eight molars, in the upper, and six in the lower jaw. They are unrooted, of simple cylindrical form, and structure. These teeth are not at all fitted for grinding. The sacculated formation of the sloth's stomach, adapted only to receive masticated, pappous shoots and leaves, makes provision for this apparent deficiency, compensating as it does for want of hardness and strength of his teeth. We accordingly find the teeth of the sloth, not at all adapted for grinding, but merely to break down the tender, downy structure of shoots and leaves.

Professor Owen perceiving the immense size of the tooth under his examination, possessing attributes for tender food only, argued, notwithstanding this apparent delicacy of structure, and feeble organization of the tooth, monstrosity in the animal owning it. This presented a difficulty. How could such an unwieldy monster as this animal must be, climb trees and shrubs, to feed and obtain subsistence, as do the sloths of our time? And again, what tree could bear such a prodigious weight? these questions had to be answered.

Professor Owen, contemplating the whole shape, and make of the mylodon, as also of the megatherium under investigation, came to the conclusion that its enormous forefeet, were formed to burrow down, and uproot trees. The animal then rearing on its hind legs, and tail, which last was of enormous dimension, by its weight pushed down the trees and felled them to the ground. The meal now lying at its feet, he browsed off its dainties at his leisure.

This doctrine as we might suppose, did not receive implicit acknowledgment, or a ready acceptance from geologists. Dr. Buckland attributes the roots of trees, as the proper food of the mylodon, arguing, that the custom of falling immense trees to gratify hunger, and sustain life, would accidently be the cause of death to the animal, necessitated to practice such means to obtain food.

The next specimen introduced from South America, helped to unravel a difficulty thus started, and somewhat reconcile these conflicting theories. It presented the discovery of *a very large fracture of the skull* of a kind which could only have occurred during life. This fact, gave a solution of the question, with respect to the probabilities of danger, to which this collossal creature was exposed, of loss of life, while practising the universal instinct, of endeavouring to sustain itself by obtaining food.

But another difficulty presented itself. It was contrary to the wise provision of the author of the universe, to leave unreconciled, and unexplained such an apparent incongruity, as coupling means of preservation, and nourishment, in combination, with almost inevitable injury and destruction.

The train of investigation of Professor Owen, in pursuing his enquiries, was necessarily directed to the developement, and formation of the skull of the Megatherium, and Mylodon.

He found the skull was remarkable, in having two tables as usual, but these tables were separated by air cells. Thus, by this provision, fracture of the outer table of the skull, would not be communicated to the inner one, or necessarily involve any injury to the brain.

Out of these slender materials of the habits, and anatomy of an extinct specimen of the animal kingdom, science has builded up a theory, which seems satisfactory, and sufficient. The teeth of the megatherium, and mylodon, all molars, consisting of four on each side in the upper jaw, and four in the lower, with a supplemental socket behind, are also especially remarkable, and distinguished for the thick coat of *cementum*, investing their body. These teeth have no roots.

Paleontological investigations have become definitely determined, as in the instance we have given above, by appeals to the comparative anatomy of the teeth of extinct, with living animals.

Thus we see the form, and structure of teeth, have assisted to determine the *genera*. and *species*, not only of existing animals, but of those which have existed, in the universe, before the creation of man. The harmonies thus observed between extinct, and living animals, in their osteological structure, and arrangement, especially in that of the economy and formation of teeth, form a basis upon which to unravel, many interesting material mysteries in nature.

The teeth therefore are not so barren an organization, as their minute structure, and low vitality would imply. On the contrary, the cursory review of the subject we have taken, establishes their claim to a position in science, and to a philosophical consideration.

CHAPTER V.

We emerge from the dark ages, to look about us in the sixteenth century, when we obtain in the midst of obscurity, and intellectual darkness, some returning light.

Eustachius, who lived in 1563, enunciated clear views of the growth, and form of the teeth. But we arrive at a much later period, before science thoroughly investigated the physiology, and economy of the human teeth. In modern times Leuwenhock first discovered the value of the microscope, having applied it in anatomical investigations, by which he discovered the tubular structure of tooth bone, or dentine. This was afterwards confirmed by Retzius of Stockholm, and Professor Owen. The science now made rapid progress, much indebted to the professors of France from an early period, amongst whom especially should be named the famous Ambrose Paré ; and in recent times the celebrated anatomist Bichat ; also the great naturalist Cuvier. Chemant, in 1789, made the discovery of artificial mineral teeth.

John Hunter, laid the foundation of the
English School of dental science, we may go
further and say with truth—he laid the foundation of the science itself. In 1771 he published
his "Natural History of the Teeth," a valuable
work, and a great accession to the art. This was
followed by many practical and scientific works
emanating from France and America, as well as
our own country, down to the present time. In
justice, we should name Blake, Fox, Bell,
Professor Owen, Nasmyth, Tomes, and Goodsir,
amongst native eminent naturalists and dentists.
Through the industrious labours and scientific
researches of these men, the dental art has arrived at its present scientific eminence.

Gardette, in France, was the originator of
the application of suction plates, on the principle
of atmospheric pressure, for support of artificial
palates and teeth. But no modern invention, or
improvement, has proved of so much value,
and importance in securing to the dental
art, the higher claim of science, as has been the
introduction of Caoutchouc. as the basis of all
mechanical procedures upon the mouth. Gardette's
suction plates, and discovery of the applicability
of atmospheric pressure, great and inestimable as
was its value, would, without the discovery of
india rubber for a basis, have been compara-
tively worthless.

It was our good fortune, to develope
this principle, when we introduced in 1852,
caoutchouc fixatures for the basis of gums,
and palates, and for support of artificial teeth in
the mouth. We successfully continue to carry out
this practice at present, having made the discovery at the very moment it was wanted. The
application, with any degree of success of atmos-

pheric pressure, for fixatures of the mouth, was altogether impossible, at the period of the art when Gardette made the discovery. For want of knowledge of a material of sufficient plasticity, levity, and resiliency, dense heavy metals, were used, which rudely, and most unsatisfactorily carried out this idea. Thus, in the absence of caoutchouc, this incomparable discovery would have proved unavailable, and have died out.

Our patent at once solved this problem, and removed difficulties, and impediments to the perfecting of fixatures for the mouth and palate. Caoutchouc, would appear to be the only material in nature, capable of mechanical manipulation, for so delicate, and tender an organ as the mouth; as atmospheric pressure would also appear to be the only cement, or tension adapted to its soft tissues.

Gutta percha, and every description of metal, had been resorted to, down to the newly discovered metal *alluminium*, with limited success, each being attended, with innumerable difficulties and drawbacks. Difficulty of adjustment, non-elasticity, or want of resiliency on the one hand; again corrosion, and early or late destruction occurs from oxidation of metals. For dentures made of metals, the hammer and file are required in their adoption, implying in the one case, necessity to use concussion, and in the other, confessing inexactness in fitting.

These difficulties are all conquered, and avoided in the use of cauotchouc. The model of the mouth being taken, the process without any further manipulation, is forthwith completed by india rubber, in a state of fusion, or sufficient liquefaction, to be enabled to receive the counter die of the mouth, in a perfect and natural state. The advantages, thus mechanically expressed, speaking eloquently in favour of our patent, are too obvious, to need further insisting upon.

Again, our patent supersedes all fixings, such as the use of ligatures, or wires of any kind, so much to be deprecated, and so injurious to the soft tissues of the mouth; they being needless, are never resorted to. Tension of teeth, by our patent is effected in most instances by natural agency, the agency of atmospheric pressure, accomplishes for it all that is required; a force equal to fifteen pounds weight of pressure, upon every square inch of free space, upon the earth's surface. This force is productive of the most wonderful phenomena in nature.

Caoutchouc has the singular virtue, of resisting the influence of all external agents. The acids of the secretions, and saliva, exercise no influence upon the material of india rubber. Thus, when every other material, used for dental fixatures, and basements, excepting gold, in a very short time becomes corroded, decomposed, and oxidized, caoutchouc remains unaltered, free from any contamination, and is as firmly fixed as, when first placed in position.

Stopping carious, and decayed teeth. is an important procedure in dental surgery: this is likewise effected by caoutchouc. Its capability of receiving a natural tint, to conceal cavities in front of the mouth, renders it superior to gold for such purpose. This virtue of india rubber, gives a supremacy also over every other material, for excluding air, from the tender nerve, and for the object of temporary stopping, until gold can be resorted to. Nothing certainly transcends, or indeed equals the quality of gold, for filling, or stopping teeth permanently.

The softness of india rubber, as a matrix, or basement; and its ready conversion into artificial gums, nearly approaching in texture nature's own alveolary tissue, prevents many accidents, so common to mineral teeth, imbedded or fastened in gold plates. This plastic virtue, renders the painful process of removing stumps, so alarming to the patient, unnecessary.

The capability of india rubber, to receive a perfect impress of the mouth, not only avoids all pressure upon existing stumps and teeth, but affords them complete security, firmness, and support; so that remaining teeth or stumps may be retained for years. Additional teeth from time to time, may also with the greatest facility and comfort be supplied upon existing basements; thus avoiding the constantly recurring expense, so much complained of with the use of gold basements.

Caoutchouc forming the frame, or basement, receives the artificial teeth in their beds, nearly as natural teeth, are received in the gums, without requiring any tube, wire, or other extraneous appliance, to secure them; and are capable from their tenuity, and lightness of any kind of adaptability. The teeth become homogeneous, or solid with the gums, being imbedded into the caoutchouc basement, while they are in that soft, semi-fluid condition, required for modelling.

Thus is obtained a combination of all required qualities, namely, of lightness, purity, and indestructibility; caoutchouc being thoroughly inaccessible, to the usual influences which prevail in the atmosphere, and in the secretions of the body. They possess also integral elasticity, so as to conform to, and assume at the usual temperature or blood heat of the human body, the natural elasticity of living tissues, and many of the contractile attributes of irritability, peculiar to organized animal matter.

Casts, from metal, necessarily undergo the bruising, and concussion of the hammer, and the manipulation of the file, to accomplish even approachment, to any degree of accuracy; both which procedures, especially that of malleation, is inconsistent with the accomplishment of the object intended, and preclude the application of our principles of atmospheric pressure.

The lightness and tenuity of caoutchouc is such, that it may be made to assume any form, with almost protean variety. Mineral teeth, a great improvement of the present day in dental art, would have had small justice meted to their undoubted advantages, and merits, but for the discovery of the applicability of suction pressure, and of india rubber fittings.

Freedom of mineral teeth from possibility of decomposition, has established their undoubted superiority. Other materials used in formation of teeth, have been found deficient in this special virtue; organic materials especially acquiring a corrupt, and disgusting odour, and taint. The capability of mineral teeth, to receive any tint, or colour desired by the wearer, and also of retaining permanently, the particular expression, or colour given, is one of their most valued characteristics. This natural appearance, allied with durability, cleanliness, and strength, commend mineral teeth to approval, possessing as they do, complete imitation of the enamel, and pellucid semiopacity of natural teeth. The only want, or requirement remaining to be supplied, was suitable basements, for fitting these newly invented dentures.

No cunning craft of the goldsmith, could
obviate the objections to metals, which we have
described. The key stone in the circle of dental
science was yet wanting, until our good fortune,
made the discovery, of the applicability of caoutchouc, and atmospheric pressure, to render the
dental art perfect by their fittings, and contrivance.

The perfection to which mineral teeth have
now attained, by the improvements of Mr. Ash,
and Mr. Lemale, have left nothing to be desired,
combining as they do, with a perfect model of the
natural tooth, its precise form, and shape, so as to
challenge comparison, and defy detection. Whereas,
natural teeth when substituted, would require a
costly annual restoration, with their fittings, and
basements, india rubber basements supplied with
mineral teeth, endure for years, or the term of
life. They may in a sense, be almost considered
superior to the original, or natural teeth, as
merely mechanical manducators, wanting only
the nervous connection conferring tactile sense,
to make them life-like. Thus, the various crafts
of graver, lapidary, and goldsmith, are engaged
in the operations of the skilful dentist, as they
were also enlisted, in the studio of Benvenuto
Cellini, to accomplish his divine achievements.

It is no common, or inferior skill that will
suffice, or screen the dentist, from severe criticism,
if he does not practice the art with skill and
judgment. Nor can the profession, with any
justice complain of want of due appreciation, by
their patrons the public. The profession is fairly
estimated in all grades of society, as capable of
conferring great benefits in regard to the health,
and comfort of individuals.

Empiricism no longer assumes pretensions, in

presence of the advanced progress of mechanical skill. The simplicity with which our caoutchouc inventions have solved every difficulty, has left no chance for mystery, or imposition. The profession is no longer an uncertain art, in which any ignorant pretender may embark, having for his motto, *fiat experimentum in corpore vili.*

> " For the dull world, most honour pay to those
> Who on their understanding most impose."

Since artificial mineral teeth, have arrived at the perfection, and purity they at present possess, they subserve every requirement dental deficiency demands. Ivory so called, produced from tusks, either of elephant, walruss, or hippopotamus, may be occasionally found advantageous, and applicable to particular circumstances.

Mineral teeth possess in a greater degree durability, strength, and incapability of corrosion or corruption, than those prepared from ivory, or even from natural teeth of any kind, human or of animals. We can not only as we have related above, imitate nature in a perfect degree so as to produce a *fac simile,* of the most beautiful type of tooth, but also give mineral teeth a permanent tint, the shade of which is so inimitably true, as to excite admiration. These are the advantages possessed by our mineral teeth, and caoutchouc basements. Mineral teeth being more durable than natural teeth; caoutchouc basements possessing also almost every quality of living organic matter; at the same time are not liable to the decomposition, and corruption inseparable from animal organic matter.

If instead of possessing so many virtues, they were accompanied with defects, and draw-

backs, this singular compensating virtue of cleanliness, in addition to purity, and incorruptibility
would give them a decided preference, and
support. In addition, they possess a recom-
mendation appreciated by all, that of economy,
and cheapness.

Few are those happy beings, who are exempt
from sufferings, which demand the regard, and
care of the dentist, we may truly say " *apparent
rari nantes*," here, and there, they may be found
amongst good, and amongst bad society, but
strange to say, are more often found amongst
the savage, than the civilized.

Teeth are lost from a multitude of accidents, and causes, yet we must enforce,
that a single tooth cannot be lost with im-
punity, to the animal economy. Animals in
a state of nature, never have their teeth diseased.
Whole tribes of savages, are exempt from dental
disease. Even our domestic animals, are less privi-
leged than the savage, since horse and dog, are
often afflicted with toothache. Yet how little regard is had to the remedy at hand, how few avail
themselves of the benefit, which at every corner
beckons them to participate. Dentists abound
almost as plentifully as parsnips ; " cur moriatur
homo, cui *salviæ* crescit in horto ?"

" If fools have ulcers, and their pride conceal 'em,
They must have ulcers still, for none can heal 'em."

Lord Chesterfield says :—" Fine and clean teeth,
are among the first recommendations, to be met
with in the common intercourse of life."

It has been our good fortune, to have practised for a cycle of ten years, with the greatest
success, and most encouraging approval of the
public, our first discovery, in the application of

caoutchouc fittings, and bases. It affords us great pleasure, not unmixed with a feeling of pride, to announce a further improvement, upon the same material. The capabilites of a new discovery are seldom tested, or only fully brought out, after experience has given its proverbial teaching. Ten years of such experience, and practise, with india rubber, have confirmed our faith in its unrivalled value, making us also familiar with its deficiencies; which we have at length succeeded in conquering in the most satisfactory manner. These deficiences, may be comprised in a few words, but the newly discovered *aurifero-vulcanite bases*, to which we now beg to call attention, entirely obviate them.

Caoutchouc fittings were accompanied with objections, most familiar to those, who least appreciated their virtues. Thus, although possessing all the good qualities, we have so much insisted on in these pages, in their manufacture, they will occasionally come from the mould with rough irregular surfaces, requiring to be polished, to prevent irritation of the sensitive tissues of the mouth. This necessary process, unfortunately disturbs, and alters the matrix of the die, preventing that exact fit, so essential to develope, the full force of atmospheric pressure.

These difficulties, and requirements, are now fully satisfied, and obtained through the agency of the *aurifero-vulcanite* envelope. By the means of this envelope, the perfect model of the mouth obtained in the mould, from the wax impress, is rendered inviolate, and permanently preserved; no mechanical interference, or disturbance being required, to remove irregularities, effect smoothness, or to give a polish.

The gold being chemically applied, neither disturbs, or in any way alters the impress, derived from the wax. Metallic dies, necessarily expand and contract, from subjection to various temperatures of heat. They from this disturbance, are rendered unfit to receive casts, of such delicate mouldings, becoming invariably more, or less warped, and strained in the process of forming them.

We have no hesitation, in announcing this improvement, since it confers so many advantages that it may be justly hailed as a new era in dental art.

We strongly advise sufferers, to resort to the early substitution of artificial teeth, upon loss, or decay of any of their natural teeth. This timely regard, prevents extension of mischief, proving the means, of affording incalculable comfort to the patient, and is also a truly economical proceeding. If neglected, the gums, and alveolary receptacles of the teeth become implicated in disease, rendering the mouth uncomfortable, by spreading mischief to the neighbouring teeth; thus, preventing the unfortunate sufferer, participating in the enjoyment of his food, so requisite, to participation in the various pleasures of life and health.

Our mineral teeth, supported on caoutchouc bases, with the valuable improvements they now possess of the aurifero-vulcanite, can be applied individually, or in sets, with every degree of capability to fulfil the functions of articulation and mastication.

CONCLUSION.

Having said this much about teeth, in many of their relations, physiological, anatomical, poetical, antiquarian, and historical, our especial object, in having so done, no doubt will be obvious to our readers. We have found this discursive treatment of teeth, an attractive, and rather fascinating theme. Teeth, have vindicated their claim to an antiquity, which no expression of lapse of time, can realize. They link the present inhabitants of the world, with a period so distant, and with inhabitants so different and remote, that probably, but few of the varieties of the present dwellers upon earth, or water, can claim any relation, or contemporaneous connection with them.

We may say that we enjoy a privilege, of somewhat ancient date in the treatment of human teeth. Our name being connected with the first patented invention for artificial teeth, in the prepared india rubber. Having in 1852, made this valuable discovery, we secured it by letters patent.

The teeth are organic portions of the animal system. Their bony substance, is much harder, than that of the bones of the body, containing less gelatinous matter. Teeth are supplied with blood vessels, and nerves, being susceptible under disease, of a high degree of sensibility.

We have before stated that enamel, may be considered extraneous to the system, not being in any manner either nourished, or reproduced. It is liable to destruction, and external injury, by the slow disintegration of chemical decomposition, or mechanical mischief, and hence offers a weak vulnerable point, for access of disease to the tooth. The bony substance of the tooth, is very liable to inflammation of its organic structure, by injury, or destruction of the enamel, or from internal sources of disease, or decay of that low kind, peculiar to bony structure. This condition of low organization, in connection with the exposed surfaces, of a great portion of the bony substance of the tooth, make it amenable to the diseases, peculiar to bone in general, as well as to other diseases, *sui generis.*

Thus, from the above circumstances, teeth do not possess sufficient vitality, to effect the process of exfoliation, of dead portions of their structure, after inflammation, as the preliminary step, to the renovation of the dead part. Other bones, are capable of this restoration, and renovation.

We do not intend to give a history of the diseases of teeth ; we may merely remark, that caries, attacking as it does the crown of the tooth, after destruction of the enamel, is the prolific agent of mischief, by ultimate exposure of the nerve to sources of external irritation. But injury, of infinitely greater magnitude, finds a mechanical remedy in our artificial india rubber palates. The victims of lost or defective palate, either from congenital malformation, or resulting from constitutional disease, are truly objects of commiseration.

Destruction of the arch of the palate, is certainly one of the greatest afflictions, to which man is amenable in social life. This condition from caries of the palatal, and nasal bones, renders the individual sufferer totally incapable of articulation, or of distinct intelligible utterance. Thus, the patient being deprived, of the most agreeable intercourse in life, becomes an alien in society. Unfortunately, this is the smallest privation to which he is subjected, by such visitation of disease. By destruction of the palatal bones, or of the vomer, and nasal bones, which are the natural divisions between nose and mouth, food and liquids, find a common way, rendering the patient an object of disgust to himself, and unfit to mingle in the social circle. Snuffling, coughing, and inarticulate attempts at utterance, aggravating the painful condition of the sufferer.

This deplorable calamity, is fortunately of rare occurrence, in the present day. Improved civilization, has brought greater attention to decency, self-respect, and cleanliness. Moreover, medical science, aided by advanced physiological investigation, has enabled us, to measure remedies, so as to appropriate, and vindicate their virtues, without approaching their vices; for in many instances, the more inveterate the poison, the more valuable is the remedy. Our India-rubber palates, supported by atmospheric pressure, are sufficient to remedy, and obviate all these distressing conditions.

The celebrated French Physician, Ambrose Paré, claims the merit of invention of artificial palates, as early as the year 1585. Ambrose Paré's invention was improved by Fauchard, and Bourdet.

Whether it be refraction, from the enamel of the teeth, which throws a playful radiation of light, over the expressive ruddy line of the lips, or not, suffice it to say, that all the delights, and passions of human nature have their expression seated on the mouth.

In the expression of anger, indignation, or disgust, the mouth becomes deformed, by elevation of the angle, showing the teeth, only in a partial, and distorted manner. In joy, and pleasure, the mouth becomes playfully relaxed, and expanded, with uniformly curling lines, displaying the teeth in smiling rows. There is great variety among different individuals in the expression of their teeth, or perhaps more properly speaking in the expression of the mouth. In the joyful, genial man, pleased with the goodness of creation, having health also to enjoy it, the mouth assumes an universal smile, and pleasure beams from his countenance.

Caoutchouc, in its improved state of manufacture by Goodyear, came to our hands, and by its capabilities, we have at length achieved, the *ne plus ultra* of mechanical invention, in artificial palates, and other fixatures, for which it is so singularly, and peculiarly adapted. Thus, articulation, mastication, and deglutition are completely restored to the suffering, and miserable patient, by this ingenious substitute, which we have invented since 1852.

Mr. Goodyear is an American, not by profession a dentist, or engaged in the art of dentistry. He greatly improved the material of india-rubber for dental, as well as for general purposes, and obtained various patents, for the improved manufacture, of vulcanized India-rubber. This gentleman, knowing our early claims in the invention, and use of India-rubber, for purposes of the mouth, on his visit to England in 1853, made very liberal overtures to us, to connect his name, as also the patents, of this new material of *vulcanized India-rubber*, to carry out our valuable discoveries.

Although fully acknowledging the merits of Mr. Goodyear's discovery, and improvements in the manufacture of India-rubber, of which we have availed ourselves, this offer was declined. We knew the original, and intrinsic merits of our system, its elasticity of application, and that it possessed such extensive capabilities, that any merely material improvements, in the manufacture, or manipulation of the staple, could be appropriated with great facility at any time.

"The divine Montaigne, who pours out all so plain," in his essays, says — "Nothing (in meats) offends me, but toughness; not for want of teeth, which I have ever had good even to excellence, and which age does but now begin to threaten; for I have ever been used every morning to rub them with a napkin, and before, and after dinner. God is favourable to those whom he makes to die by degrees; 'tis the only benefit of old age; the last death will be so much the less full and painful; it will kill but half, or a quarter of a man. I had one tooth lately fall out without drawing, and without pain; it was the natural term of its duration; both that part of my being, and several others are already dead, and others half dead, of those that were most active, and in highest esteem during my vigorous years; so methinks I melt, and steal away from myself."

Thus we can easily conceive the anxiety manifested by well conditioned people for the state of their teeth, our wonder being that the feeling is not more universal.

The human teeth, subserve to the purpose of speech, articulation, and voice, as well as for preparing food, for the process of digestion. This is not the case in other animals than man, When in advanced age, teeth are lost, and alveolary processes become absorbed, man is no longer a speaking, feeding, or peptic animal. Life, at this time is preparing for that re-newed juvenility, only to be arrived at and consummated through death, decay, and cor-ruption—an universal penalty inflicted by the laws of life, to fulfil the ordinations of the universe. Thus speech is lost when teeth are lost,

"Nec vox hominem sonat."—*Virg.*

John Hunter says, of the front teeth, "They come on the child, just as he is beginning to speak, and at that time, are so loose, as really to be of little use in masti-cation," which circumstances would imply, that their immediate function, was to develope the voice, and assist articulation, and speech in childhood.

In all the races of mankind, the teeth are found similar ; the number and conform-ation, being the same amongst the various races of man.

Among those tribes, in whom the jaws are prominent, namely, the Negro, and also in the Guanche mummies, of the Canaries, as noticed by Dr. Prichard, the front incisors, have an oblique or slanting position, being also thick in their bodies; thus, instead of having the ordinary cutting edges, they have the appearance of truncated cones. The same peculiarity has been noticed in Egyptian skulls. It is doubtful whether this is a normal, or original difference in form, or not.

The direct, or perpendicular facial line, is not seen in other animals, than man. The facial line, in some animals, ranging from an oblique, almost approaches the horizontal line. The small pouting mouth, with erectile lips, round, and prominent cheeks, nose smoothly developed, with uniform teeth, characterize the physiognomy of man. The vertical position of the incisors, in the lower jaw of man, assist to give that remarkable prominence of chin, terminating in the almost square maxillary base, entirely absent in animals, whose nose, mouth, and lower jaw, are nearly on a parallel line.

Next to the eye, no doubt the mouth is the seat of most expression in the human face. The face thus beautifully developed, becomes the index of the mind—the stage upon which the passions play, and from which we may derive our knowledge of the inner man.

www.ingramcontent.com/pod-product-compliance
Lightning Source LLC
Chambersburg PA
CBHW022010190326
41519CB00010B/1471